Best Manage

Managing Trees During Site Development and Construction

Third Edition

Nelda Matheny
E. Thomas Smiley, PhD
Ryan Gilpin
Richard Hauer, PhD

ISBN: 978-1-943378-28-9

Copyright © 2008, 2016, 2023 by International Society of Arboriculture
All rights reserved.

Printed in the United States of America.

Except as permitted under the United States Copyright Act of 1976, no part of this publication may be reproduced or distributed in any form or by any means or stored in a database or retrieval system without the prior written permission of the International Society of Arboriculture (ISA).

Standards and regulations such as ANSI, OSHA, ISO, and others regionally and internationally may have specific requirements for use of "shall," "should," and "must." For the sake of readability and to account for differences between standards, this publication does not necessarily parallel the requirements of such standards or regulations in the use of these terms.

Cover Design: Veridian Design Group
Cover Photo: E. Thomas Smiley
Page Design and Composition: Veridian Design Group
Illustrations: Bryan Kotwica, Taylor Design (Figure 9 TPZ Fencing, Figure 10), and
Nelda Matheny (Figure 9 Plan Detail)
Printed by Premier Print Group, Champaign, IL

International Society of Arboriculture
270 Peachtree St. NW, Suite 1900
Atlanta, GA 30303
United States of America

Gray's Inn House
127 Clerkenwell Road
London EC1R 5DB
United Kingdom

+1.678.367.0981
www.isa-arbor.com
permissions@isa-arbor.com

10 9 8 7 6 5 4 3 2
1123-CA-1300

Contents

Acknowledgments

The *Managing Trees During Site Development and Construction* Best Management Practices (Third Edition) Review Committee:

- Don Zimar, Manassas, VA
- Mark Williams, City of Fort Lauderdale, Fort Lauderdale, FL
- Christopher Riley, PhD, Bartlett Tree Research Laboratories, Washington, DC
- Ellen Vincent, Clemson University, Clemson, SC
- Rick Till, Honl Tree Care, Portland, OR
- Paul D. Ries, College of Forestry, Oregon State University, Corvallis, OR
- Lindsey Purcell, Indiana Arborist Association, Franklin, IN
- Andrew S. Potter, Principal Arboricultural Consultant, Potter Tree Consultancy, Boat of Garten, Scotland, United Kingdom
- Craig Pinkalla, Minneapolis Park & Recreation Board, Minneapolis, MN
- Kaitlyn Pike, University of British Columbia, Vancouver, Canada
- Johan Östberg, Director, Nature Based Solution Institute, Sweden
- Tom Mugridge, Forest City Tree Protection Co., Mayfield Village, OH
- Verna Mumby, Mumby's Arboriculture Consulting, British Columbia, Canada
- Sharon J. Lilly, Acworth, GA
- Colin McFeron, City of Shoreline Public Works, Shoreline, WA
- Gordon Mann, California Tree and Landscape Consulting, Inc., Auburn, CA
- Jamie Lim, ISA Board Certified Master Arborist®, New York, NY
- Janine Lester, Landscape Design Consulting, Richmond, VA
- Cene Ketcham, Wetland Studies and Solutions, Inc., Millersville, MD
- Kent E. Holm, Douglas County Environmental Services, Omaha, NE
- Mark J. Hoenigman, Busy Bee Services, Ltd., Novelty, OH
- Timothy Dickinson, PLC, Wood, Tecumseh, Canada
- Scott Diffenderfer, Town of Vienna, VA
- Andrew Driscoll, Montgomery Parks, Gaithersburg, MD

Susan D. Day, University of British Columbia, Vancouver, Canada

Terrence P. Flanagan, Teragan & Associates, Inc., Lake Oswego, OR

Steven Geist, SavATree Consulting Group, Aurora, CO

Richard Gessner, Monarch Consulting Arborists LLC, Felton, CA

Jason Hasaka, Principal Consultant, Bartlett Tree Experts UK & Ireland, Bristol, United Kingdom

Beth Brantley, PhD Bartlett Tree Research Laboratories, Chambersburg, PA

Jerry Bond, Urban Forest Analytics LLC, Geneva, NY

Kristina Bezanson, University of Massachusetts Amherst, Amherst, MA

Andrew Benson, PhD, The Tree Consultancy Company, Auckland, New Zealand

Maryellen Bell, Bartlett Tree Experts, Austin, TX

Camilo Ordonez Barona, University of Toronto, Toronto, Canada

Scott D. Baker, Tree Solutions Inc., Seattle, WA

Purpose

The International Society of Arboriculture (ISA) has developed a series of Best Management Practices (BMPs) for the purpose of interpreting national tree care standards and providing guidelines of best practice for arborists, tree workers, and the people who employ their services.

This BMP is intended to help owners, developers, arborists, and urban foresters preserve and manage trees during development. Arborists should be involved during all phases of a development project: planning, design, preconstruction, construction, landscape, and postconstruction. The arborist's involvement may be as a consultant or as a municipal, utility, or commercial arborist.

This BMP was created as a companion publication to the *American National Standard for Tree Care Operations—Tree, Shrub, and Other Woody Plant Management—Standard Practices (Management of Trees and Shrubs During Site Planning, Site Development, and Construction)* (ANSI A300, Part 5), but is also intended to support standards from other countries.

Because trees are unique living organisms, not all practices will be relevant to every situation. A qualified arborist should write or review contracts, specifications, and reports utilizing applicable standards and this BMP. Departures from the standards should be made with careful consideration of the objectives and with supporting rationale.

Arborists and their clients should be aware that even when the BMPs are applied diligently, trees may decline or die, either as a result of the construction impacts or from other causes outside their control. There is always a chance when building near trees that a tree may not survive, but the potential for successful tree preservation is enhanced with diligent adherence to standards and this BMP.

1. Introduction

Construction or renovation of buildings, utilities, roads, parking lots, and other structures occurs as communities grow and land use changes. Trees and other vegetation on construction sites may be relocated, removed, or preserved depending on regulations, development goals, and project specifics. The principles and processes described in this BMP can be applied to preserve soil, shrubs, and other woody vegetation, alone or in conjunction with tree preservation.

Preserving trees is important to sustain the cumulative environmental, social, and economic benefits they provide. Trees improve air quality; conserve energy; reduce stormwater runoff; sequester carbon; help connect people to nature; and, more broadly, help create livable, healthy communities. When mature and established trees are removed, many of these benefits are lost. Newly planted trees often take decades to provide similar benefits.

Why Should Trees Be Preserved on a Building Site?

- Legal Protection—Tree preservation may be required to comply with legal requirements in the form of deed restrictions, ordinances, or regulations. There may also be government or private sector (e.g., SITES) incentives for preserving trees.
- Social and Cultural—Trees' aesthetics define the character of a place, can help stimulate recreation and social interactions, and have historical or cultural significance linking people to the place or past events. Trees provide physical and mental health benefits by promoting physical activity and a positive mental state.
- Environment—Trees provide benefits through reduced heating and cooling costs, increased carbon sequestration, stormwater management, wildlife benefits, air purification, and noise attenuation.
- Monetary Value—Trees typically increase property value and improve property sales. Mature trees can improve the sense of space and curb appeal of a project, home values, home sales, and occupancy of leased facilities. People enjoy living, shopping, and working near trees.

Government regulations or incentives may exist to encourage or require tree preservation during site development and construction. In some situations, developers and property owners proactively identify trees to preserve because of the values they bring to their property or project. Arborists can have a key role in development projects by evaluating the suitability of trees for preservation and by providing appropriate protection specifications and strategies.

Preserving trees during a construction project generally involves protecting the soil, roots, trunks, and crowns of trees. Projects that attempt to preserve trees without protection reduce the likelihood of long-term tree survival.

Successful tree preservation also depends on the commitment of the property owner or developer to modify designs as appropriate to protect trees and to ensure all contractors adhere to the tree preservation specifications. Without their support and enforcement, the efforts of the arborist may not be successful.

Arborists' roles in construction projects may include:

- Evaluating the feasibility of a project in relation to trees on the site
- Assessing tree suitability for preservation
- Evaluating whether trees are likely to survive planned construction
- Assisting the owner and/or development team with site layout and design
- Determining appropriate tree protection zones and providing specifications for tree protection measures
- Complying with or enforcing permitting agency requirements
- Monitoring trees and protection measures
- Implementing tree care throughout the process

Successful tree preservation is measured over the long term when trees continue to thrive after development. Well-intentioned land development projects may inadvertently damage trees on construction sites or adjacent properties. If tree health and stability are compromised due to this damage, the trees may need to be removed, often at an increased cost. In addition, years of investment in tree growth and benefits will be lost.

Key Terms

There are several terms in common use that describe the process of preserving existing trees on sites during development and construction. In this document, we use **tree preservation** to refer to the process of protecting trees and other vegetation, including the soil in which they grow, from damage related to development and construction activity as well as planning for their long-term health and stability. Tree preservation is synonymous with **tree retention**, commonly used in the United Kingdom.

Trees to be preserved should have protection. **Tree protection** is the process of excluding activities that could cause damage to a tree or soil that would negatively affect tree health or longevity.

A **preserved tree** is one that is or was protected from development and construction-related damage.

A **protected tree** is a tree that, because of its size, species, age, or other qualities, requires permission from the permitting agency to remove, damage, prune, or otherwise impact.

The **tree protection zone (TPZ)** is the defined area within which certain activities are prohibited or restricted to prevent or minimize potential injury to designated trees.

When trees are considered in all phases of a construction project and managed as outlined in this publication, most trees designated for preservation will survive and flourish.

Site development typically occurs in six phases: planning, design, preconstruction, construction, landscaping, and postconstruction (Figure 1). For a small residential remodel project, the phases may be compressed into a short time frame, while phases for large projects could extend to months or years.

Failing to consider trees during any of these phases can result in the compromise of tree health and structural integrity (Figure 2). Arborists may be involved at any phase. The greatest success is achieved when arborists are brought in early in the development process and are involved in all phases of the project.

Development Phase	Arborist Involvement
Planning	Resource evaluation Permitting needs Suitability for preservation Tree inventory
Design	Tree Impact assessment Tree protection plan Tree Protection Zones (TPZ) Landscape plan review
Preconstruction	Contractor communication TPZ barrier installation Aboricultural treatments
Construction	Site monitoring Assessing impacts Maintaining TPZ Arboricultural treatments
Landscaping	Site monitoring TPZ barrier adjustments Mitigate tree impacts
Postconstruction	Site monitoring TPZ barrier removal Mitigate tree impacts Plan for maintenance

Figure 1. Arborist involvement in different phases of a construction project.

Scope of Work

The scope of work should be defined before any work begins. It should include:

- Phases of development in which the arborist will be involved
- Arborist responsibilities at each phase
- Location/area to be included
- Data to collect during tree and site evaluations
- Schedule or frequency of site visits
- Type, timing, content, and recipient of reports

Figure 2. The differences in tree health between adequately protected trees (left) and trees without protection (right) after construction. Trees with protection are more likely to remain healthy, while trees lacking protection are more likely to decline or die and need replacement.

Objectives

Objectives of managing trees and other vegetation on a construction site include:

- Protecting trees, shrubs, other plants, and soil to maintain plant health during and after construction
- Complying with regulations regarding tree preservation on construction sites
- Determining the tree resources present on a development site
- Minimizing conflicts between trees and new infrastructure
- Developing a plan to minimize damage to the trees, shrubs, other plants, and soil
- Integrating existing trees with new landscape
- Developing postconstruction landscape management plans

2. Planning Phase

During the planning phase, the property owner and developer begin to assemble a team to perform a feasibility study that explores potential land uses and considers possible site layouts. Tree preservation is most successful when the arborist is part of the project planning team to provide information and recommendations about the trees on and adjacent to the site, as well as protection requirements and the trees' suitability for preservation. The focus of the arborist at this phase is to gather information about the project, site, and permitting agency requirements to complete a resource evaluation. The information the arborist collects during the planning phase should be used in the design phase to locate the building or infrastructure where the least impact on trees will occur and the most desirable trees will be preserved.

Resource Evaluation

A resource evaluation is the assessment of trees or other vegetation, and/or soil, present on or near the site, and the determination of legal requirements, potential challenges, and opportunities. The scope of the resource evaluation should be defined by the project and any legal requirements.

Resource Evaluations

A resource evaluation is the assessment of trees, other vegetation, and/or soil present on or near the site, as well as the determining of legal requirements, potential challenges, and opportunities. They typically include:

- Determining and communicating permitting agency requirements for resource evaluation, reporting, preservation, or other factors that affect construction activities
- Locating and describing individual or groups of trees, shrubs, and/or other woody plants on and closely adjacent to the site
- Observing the soils, hydrology, and topography in which the vegetation is growing
- Inventorying trees and other specified plants to determine location, species, size, condition, and other factors

Legal Requirements for Tree Preservation

In addition to issuing building permits, municipalities or other permitting agencies may have requirements for tree preservation during development. These requirements may define which trees are protected, often based on species, size, age, location, and/or other qualities. Protected status usually indicates that a permit is required to remove, prune, damage, or otherwise impact the tree. Application for a permit may require a report prepared by an arborist and descriptions of the trees and construction activities planned.

Regulations often define a particular protected status for trees that may vary depending on their location or definition. For example, public trees located along streets or in parks could be protected regardless of species or size while those on private property may be subject to different regulations. Often, protected trees are defined as heritage, landmark, ancestral, historic, significant, street, park, regulated, or other terms conveying their importance to the community. Arborists need to know the regulations that are applicable to the construction site and understand how to help the project comply. There is an array of potential private property deed restrictions or contractual agreements that may require preservation of trees during development. These may be in the form of deed restrictions; covenants, conditions, and restrictions (CC&Rs); access easements; view easements; or conservation easements.

Trees commonly grow across property lines. When it is likely that construction will impact a neighbor's tree crown or roots, the arborist should consider these trees in the resource evaluation/tree inventory and when developing tree protection zones. Neighbor tree conflicts can potentially be avoided by considerate planning and guidance from the project arborist.

Tree Inventory

The resource evaluation commonly consists of a tree inventory.[1] Inventories can provide different amounts of information on existing trees (Figure 3). Results of the inventory can help guide recommendations made in future phases.

For large sites, delineation of groups of trees may be the best option. Such an analysis provides a broad overview of the site vegetation to identify opportunities and constraints for developing conceptual plans. Vegetation is

1 *See* Bond 2013.

Construction-Related Tree Inventory

Scope of Work Specifications

Specifications for a construction-related tree inventory should include:

- Area to be assessed/inventoried
- Resources to be inventoried (e.g., trees, shrubs, soil)
- Minimum size tree to include (e.g., 6 in [15 cm] minimum trunk diameter or 20 ft [6 m] minimum height) based on local regulations or project specifics
- Tree inventory data to be collected (see below)
- If a tree risk assessment or likelihood of failure will be assessed, the methodology shall be specified (e.g., ISA BMP risk assessment methodology)
- Time frame for data collection (e.g., inventory completion date)
- Type of report to be provided (e.g., oral, written, map, plan)
- Due date of report
- To whom the report should be submitted

Tree inventory data should include:

- Tree location (map reference or coordinates)
- Identification/tag number
- Species (common name or scientific name)
- Size (e.g., trunk diameter, height, crown spread)
- Condition (i.e., health, structure, form)
- Suitability for preservation
- Additional factors as specified

mapped and species components are described along with open spaces. This information can be collected from a ground survey, remote sensing technology (e.g., satellite imagery, aerial photographs, lidar), or a combination of these methods. This may be used to guide initial discussions on which groups of trees may be most desirable for preservation.

Figure 3. Tree inventory can be a major component of the resource evaluation.

A tree inventory provides more detailed information. Data collected in an inventory varies based on the project needs. It should include all the information required to make appropriate recommendations and meet permitting agency requirements. Inventory data to be collected should be defined in the scope of work.

Location

The spatial location of trees relative to other design components is essential to tree preservation. Arborists may locate trees with an accuracy of a few feet (to within one meter) using global positioning systems (GPS) in conjunction with geographic information systems (GIS). However, most tree preservation projects require a higher level of precision (e.g., within a few inches or centimeters), and many permitting agencies require a professionally licensed or certified surveyor. A professional land survey typically plots the location of each tree's trunk and sometimes crown dimensions or other tree features.

In addition to determining the location of a tree, the arborist should consider attaching a unique identification tag to trees so they can be referenced in project plans, tree inventory data, and reports. Physical tags on each tree can reduce confusion when construction workers and arborists undertake work near or on trees.

Species

During the inventory or resource evaluation, individual or groups of trees should be identified to the species level. Species identification is important to assess construction tolerance, maintenance needs, and pest issues that may affect tree preservation. In some locations, certain species are protected, while others are considered undesirable or not required for preservation.

Size

Tree size can be important for determining protected status and the area for the tree protection zone (TPZ). Tree size is usually reported as the trunk diameter at breast height (DBH). In the United States, the DBH is typically measured at 4.5 ft (1.37 m) above grade. The diameter measurement should consider lean, slope, multiple stems, and other factors.[2] Other countries and

---

2 *See* Roman et al. 2020.

some permitting agencies have different standard measurement heights or ask for different measurements (e.g., British Standard BS 5837).

Common measurement tools include diameter tape, Biltmore stick, calipers, and visual estimates. The arborist should describe the methods and level of precision used in inventory.

Tree height and crown spread (drip line) measurements may also be helpful and required by some permitting agencies. Tree canopy diameter or radius dimensions are measured in one or more directions. These measurements may be helpful for determining TPZs as well as the influence a tree crown will have on building placement, road location, and the space where sunlight will be desirable. Crown dimensions may also inform pruning needs near the proposed building, as well as canopy replacement values.

Tree height and crown spread can be measured by the surveyor or by the arborist by pacing the distance or by using a tape measure, measuring wheel, or laser range finder. Tree height is commonly measured with a clinometer or hypsometer.

Tree Condition

Tree condition can be assessed in a variety of ways, depending upon the project. Condition can be composed of three distinct but often related qualities: health, structure, and form.[3] The relevant parts of condition are ones that relate to the decisions being made about preservation.

Tree health (vigor) includes evaluation of crown density, foliage color, leaf size, annual shoot growth, disease infections, insect infestations, the presence of injuries, and shoot dieback symptoms.

Tree structural evaluation includes assessing tree features impacting structural integrity. These conditions may include the presence of codominant stems, weak branch attachments, decay indicators, root collar burial, dead or dying branches, low live crown ratio, or other conditions of concern. The structural evaluation may include rating the likelihood of failure. This information could then be used to rate tree risk and determine risk mitigation treatments when

3 See Council of Tree and Landscape Appraisers 2019.

the site plans are developed in the design phase and tree targets are known.[4] If a tree risk or likelihood of failure assessment is desired, it must be specified in the scope of work.

Tree form considers the tree's growth habit—its symmetry, shape, or silhouette. Form is based on tree species (genetics), site conditions, proximity to other trees or structures, and management history.

Condition ratings may be categorical (good, fair, poor), numeric (0 to 5), or a percentage (0 to 100 percent). These may be separate factors (i.e., health, structure, and form) or combined into a single rating. The arborist should describe the condition rating system and level of precision used.

Age and Potential Longevity

Trees are often categorized by life stage as young, semimature, mature, and old. Young and semimature trees are often better able to respond to site changes and injuries than older trees, and the benefits they provide to the site and wider landscape can persist beyond the life expectancy of older trees. As trees grow larger, benefits increase significantly. Older trees can be preserved on development sites but must be provided adequate space for root and crown protection. Where an old tree will be preserved, the arborist should discuss with their client the benefits it will provide and likelihood of structural failure.

Suitability for Preservation

Suitability for preservation is a categorization of a tree's potential to be an asset to the project following development. While it is future focused, ratings of suitability for preservation are based on the species, current size, current condition, and species tolerance to construction. It is not based on specific construction plans or anticipated impacts to the tree, which may be unknown in the planning phase.

The ability of a tree to tolerate construction-related changes varies greatly among species (Table 1). The arborist's knowledge of species response to construction impacts, such as root and crown removal or changes in exposure and/or site conditions, should be employed when assessing suitability for

4 See Smiley et al. 2017.

Table 1. Examples of species tolerance to construction damage. Note that tolerance ratings may vary with geographic location (adapted from Matheny and Clark 2023).

Common Name	Scientific Name	Tolerance to Construction	Comments
Sycamore maple	*Acer pseudoplatanus*	Medium	
Red maple	*Acer rubrum*	High	Response probably associated with geographic location. Tolerant of root pruning and saturated soils.
European birch	*Betula pendula*	Low	
Atlas cedar	*Cedrus atlantica*	Medium	
European hackberry	*Celtis australis*	Medium	Needs irrigation postimpact.
Eastern redbud	*Cercis canadensis*	Medium	Response constrained by soil aeration and availability.
Camphor	*Cinnamomum camphora*	Medium	Avoid fill, wet soils. Best with irrigation after impact.
River red gum	*Eucalyptus camaldulensis*	High	Tolerant of fill and wet soils.
European beech	*Fagus sylvatica*	Low	Mature trees particularly susceptible.
White ash	*Fraxinus americana*	Low	Tolerant of root loss. Intermediate in tolerance to saturated soils. Low of mechanical injury (poor compartmentalization). Response constrained by soil and water availability.
Ginkgo	*Ginkgo biloba*	High	Tolerant of root pruning.
Thornless honeylocust	*Gleditsia triacanthos* f. *inermis*	High	Tolerant of root pruning and site disturbance. Intermediate in tolerance to saturated soils.
Kentucky coffeetree	*Gymnocladus dioicus*	Medium to High	Intermediate in tolerance to root loss and saturated soils. Tolerant of site disturbance.
Monterey cypress	*Hesperocyparis macrocarpa*	Low	Low of site disturbance.
Black walnut	*Juglans nigra*	Low to Medium	Low of root loss. Intermediate in tolerance to saturated soils. Low of mechanical injury (poor compartmentalization). Response constrained by soil aeration and water availability.
Black gum	*Nyssa sylvatica*	High	Response constrained by soil aeration and water availability.
Canary Island date palm	*Phoenix canariensis*	High	
Norway spruce	*Picea abies*	Low to Medium	Often windthrows. Low of root loss.
London plane	*Platanus* x *hispanica*	Low to High	Response appears to be location dependent. In western United States, tolerant. In eastern United States, stress intolerant in northern part of range. Best with irrigation after impact.
Flowering cherry	*Prunus serrulata*	Medium	Best with irrigation after impact.
Pin oak	*Quercus palustris*	Medium to High	Intermediate in tolerance of root loss and saturated soils.
Japanese zelkova	*Zelkova serrata*	Medium	Best with irrigation after impact.

preservation. Despite species tolerance guidelines, individual plant response can vary from the species norm depending on factors such as tree genetics, tree health, root distribution, soil volume, pests, previous injuries, geographic region, and postconstruction maintenance.

Species-related qualities may be included in the suitability rating. For example, species with a short life span or a nuisance pest problem have a lower rating. Locally invasive species may be rated lower if the species is likely to degrade nearby ecosystems.

In general, trees with high suitability for preservation are in good condition, have long remaining life spans, are desirable, and are species that tolerate construction damage. Trees with low suitability for preservation include those that are in poor condition, have short remaining life spans, have poor aesthetics, are intolerant of construction damage, or are invasive. Trees with moderate suitability for preservation are in between these two categories. They may have conditions or qualities that could be mitigated with arboricultural treatments such as pruning, pest management, soil management, or supplemental irrigation.

Suitability for preservation can be rated categorically (low, medium, high), numerically (0 to 5), or as a percentage (0 to 100 percent).

Tree Appraisal, Construction Bonding and Deposits

Some permitting agencies require a bond or monetary deposit to be used in case of tree injury or death. There are several approaches that can be used to determine the preconstruction value. Some jurisdictions specify the appraisal technique to apply.

A statutory value is an appraisal approach in which a permitting agency defines how the tree is appraised. Often this is based on trunk diameter, circumference, or other measurement. Typically, with statutory values, the DBH measurement is multiplied by a fixed dollar value to calculate the tree value (e.g., if trees are assigned a statutory value of $100 per diameter unit and the tree is 30 units in DBH, the value would be $3,000).

When a statutory value is not applied, the methods outlined in the Council of Tree and Landscape Appraisers' (CTLA) *Guide for Plant Appraisal* provides information on how to perform appraisals and what approaches may be appropriate for the defined appraisal problem. The specific approach will depend on the purpose and use of the appraisal. For purposes of bonding, the Cost Approach, Trunk Formula Technique is typically applied.

Soil, Topography, and Hydrology

Soil, topography, and hydrology dictate where and how roots have developed on existing trees. Soil information will help the arborist evaluate potential impacts to trees and develop specifications for TPZs and management.[5] The site engineer collects soil information to determine the site work requirements to construct stable structures such as soil compaction requirements, foundation design, construction method and depth, and pavement sections. The information in this report may help the arborist understand important soil characteristics. The arborist should consider soil profiles, texture, organic content, and groundwater level.

Reporting During the Planning Phase

Information gathered by the arborist should be used to assist the development or design team as they decide where buildings, roads, and other infrastructure will be placed. To do this, the arborist should provide a report to the client containing tree information gathered over the course of the planning phase.

The report may be for an individual tree on a small residential site or for thousands of trees on hundreds of acres (hectares) in a large development project.

The report typically includes:

- a description of the tree resource or tree inventory data,
- a tree location map or plan (may be developed by others on the team),
- a list of trees that warrant special consideration for preservation based on size, species, and/or other attributes,
- a list of trees that need further inspection to assess their structural condition or suitability for preservation, and
- the TPZ (not always included).

5 See Scharenbroch and Smiley 2021.

3. Design Phase

During the design phase, the location, size, and shape of the proposed buildings, landscape, utilities, and other infrastructure are decided and plotted on a site plan. Project design is typically a dynamic, collaborative effort, and ideally the arborist should be a part of the design team to address tree-related issues along with the developer/owners, planners, civil engineers, architects, and landscape architects.

Using the site plans and knowledge of the tree resource, the arborist evaluates which trees can be preserved and which should be removed or transplanted. The arborist then recommends a TPZ for those to be preserved. At this stage of the process, the arborist may also suggest modifications to the proposed design and site layout, as well as alternative building methods, to minimize root loss and other impacts to protect trees and soils.

If a tree inventory was not completed in the planning phase or was limited to basic tree information (e.g., tree stand delineations or limited inventories), the arborist should either complete a tree inventory or update the existing information to the level required. It may be necessary to reexamine trees that will be close to the construction activities to more precisely define the TPZ, evaluate potential risk, and/or determine tree work needed prior to construction.

Developers, owners, and permitting agencies sometimes choose, or are required, to preserve or relocate trees with undesirable characteristics. The arborist should discuss with the client potential implications of these actions and recommend either remedial treatments or, if there are none, tree removal. Where an otherwise unsuitable tree must be preserved, the arborist should discuss with the client the expected longevity, consequences of structural failure, remedial treatments, and ongoing maintenance requirements associated with tree retention.

Defining the Tree Protection Zone

Early in the design phase, a key task for the arborist is to determine the amount of space needed for each tree to be preserved and communicate that to the design team. The tree protection zone (TPZ) is the defined area within which certain activities are prohibited or restricted to prevent or minimize potential injury to designated trees. The size and shape of a TPZ should consider tree species response to construction impacts, size, condition, and maturity, in addition to the location of current infrastructure, planned construction, and specific aspects of the site.

The spread of root systems is often asymmetrical, especially where trees have grown next to streets, buildings, and other structures, or in layered, rocky, and/or restrictive soils. The shape of the TPZ should be adjusted to account for asymmetric root zones. Root inspection excavations[6] can help to determine the presence, size, quantity, and depth of roots. Various tools can be used for excavation, including shovels and other small tools to excavate by hand, air excavation tools, hydro excavators, or mechanical excavators with qualified/supervised operators.

Arborists should recommend a TPZ that is large enough to maintain the health of the tree without needlessly restricting construction. There are several methods for estimating an appropriate TPZ (Matheny and Clark 2023). Some permitting agencies may direct how TPZs should be determined within their jurisdiction.

Defining the TPZ is often a two-step process. The arborist often starts with a calculated TPZ. As the project progresses, the size and shape of the TPZ may need to be adjusted to identify the specified TPZ, where the TPZ barrier will be located.

A TPZ can be calculated for trees in good condition using the trunk diameter, the species tolerance to construction, and the age-class of the tree (Table 2). With this method, the radius of the TPZ is calculated by multiplying the DBH by a factor of 6 to 18, depending on the tree age and species construction tolerance. The unit of measure used to calculate the DBH (typically inches or centimeters) will be the same unit as the radius of the TPZ. For example, a mature tree with medium tolerance to construction damage (a factor of 12) and a 30 in (76 cm) trunk diameter would have a TPZ radius of 360 in, which is equal to 30 ft (9 m). For trees with DBH less than 10 in (25.4 cm), a minimum of 5 ft (1.5 m) radius TPZ should be considered. This document will refer to this as the calculated TPZ. This calculation should be adjusted for very large or small trees. For example, when the minimum calculated TPZ is less than 5 ft (1.5 m) in radius, a 5 ft (1.5 m) radius should be recommended rather than the smaller size.

---

6 *See* Costello et al. 2017.

Table 2. Guidelines for calculating tree protection zone radius for trees in good condition (adapted from Matheny and Clark 2023). To use this table, determine the TPZ multiplication factor based on the species tolerance to construction damage and relative tree age. That number (TPZ multiplication factor) is then multiplied by the tree trunk diameter (DBH). The result is the radius of the TPZ in the same units used to measure DBH. That number is usually converted to feet or meters.

Species Tolerance to Construction Damage	Relative Tree Age*	Multiplication Factor
High	Young or semimature	6
	Mature	8
	Old	12
Medium	Young or semimature	8
	Mature	12
	Old	15
Low	Young or semimature	12
	Mature	15
	Old	18

*Young to semimature = less than 40 percent life expectancy; mature = 40 to 80 percent life expectancy; old = greater than 80 percent life expectancy

The calculated TPZ is most helpful to the design team at the end of the planning phase or at the beginning of the design phase when use of space is being determined. The design team can use the calculated TPZ to plan for adequate space to preserve the trees (Figure 4). Calculated, circular TPZs are a good starting point, and some projects will use these calculated TPZs throughout the project. However, in practice, tree protection barrier configurations that follow the limits-of-disturbance and preexisting structures are more common than circular barriers.

As the design phase progresses, more in-depth decisions about whether trees are likely to survive construction and the best way to protect them may be required. The arborist may need to adjust the calculated TPZ depending on site-specific conditions to define the location of the specified TPZ.

The arborist determines the specified TPZ by evaluating on-site conditions, orientation of the canopy and visible roots of the specific tree, and planned construction. As an example, the specified TPZ may be increased in one or more directions to protect a low scaffold branch that extends beyond the calculated TPZ (Figure 5). Reducing the size of the specified TPZ on one or more sides of the tree may be necessary if planned construction cannot be

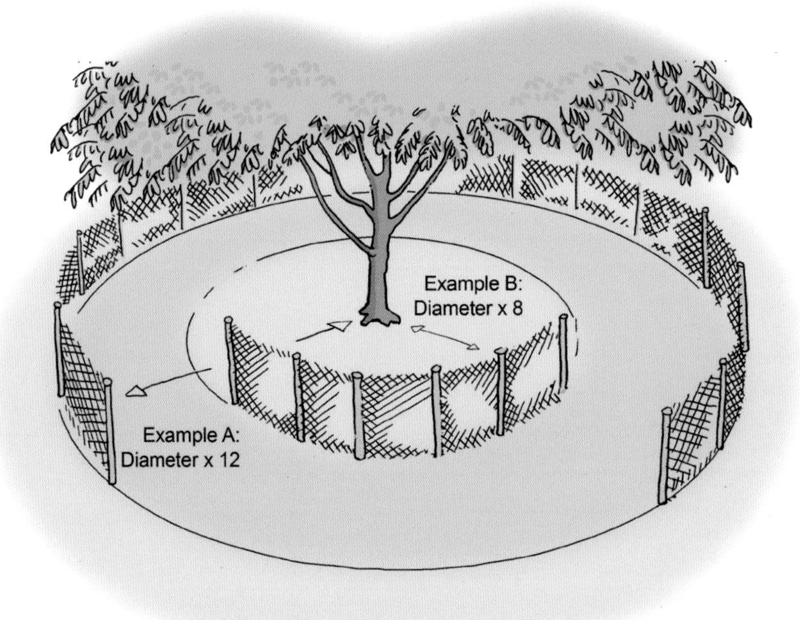

Figure 4. The calculated tree protection zone (TPZ) based on tree diameter, species tolerance to construction damage, and maturity. Applying the trunk formula method of determining the calculated TPZ by multiplying DBH × TPZ multiplication factor = TPZ radius.

Example A: Calculated TPZ for a medium-tolerance, mature tree with a diameter of 20 in (51 cm). DBH × 12 = 20 in (51 cm) × 12 = 240 in (610 cm) = 20 ft (6 m) radius.

Example B: Calculated TPZ for a high-tolerance, mature tree with a diameter of 20 in (51 cm). DBH × 8 = 20 in (51 cm) × 8 = 160 in (406 cm) = 13.3 ft (4.1 m) radius.

located outside the calculated distance. In some cases, increasing the specified TPZ on the opposite side of the tree may help compensate for roots lost elsewhere. The specified TPZ should not be reduced to the point where serious tree damage is expected.

If trees on adjacent properties have branches and roots extending into the construction area, TPZs should be established on the project site. Tree protection measures should be similar to those for on-site trees.

In addition to preserving existing trees, additional areas for root growth or tree planting should be preserved in a protected, undisturbed state, if possible. The goal is to protect the soil structure and resources so trees will thrive.

Figure 5. Calculated vs. specified tree protection zone. The calculated TPZ is determined and plotted as a circle with a defined radius. A specified TPZ is adjusted to accommodate existing and future site conditions and activities, as well as tree canopy and root characteristics, and usually is irregular in shape.

There is a limit to the amount of root or crown loss that a tree can tolerate regardless of postconstruction arboriculture treatment. If many roots, large roots close to the trunk, or large portions of the crown are removed, it may be necessary to recommend tree removal. The arborist should assess whether the tree protection zone adequately protects the critical root zone (CRZ).

Preserving Groups of Trees

Where trees are growing in a stand or are closely spaced, trees share root space and shelter each other from the wind. In this situation, it is often preferable to preserve groups rather than individual trees. Preserving a mix of tree ages, species, and understory plants within a stand will conserve a natural assemblage and reduce exposure of interior vegetation.

Definitions of the Tree Protection Zone Terms Used in This Document

critical root zone (CRZ)—area of soil around a tree where the minimum volume of roots considered critical to the structural stability or health of the tree are located. There are no universally accepted methods to calculate the CRZ.

tree protection zone (TPZ)—defined area within which certain activities are prohibited or restricted to prevent or minimize potential injury to designated trees, especially during construction or development. The TPZ should encompass the CRZ, based on the judgment of the arborist.

calculated tree protection zone (calculated TPZ)—a TPZ that is calculated using the trunk diameter and a multiplication factor based on the species tolerance to construction and age of the tree. It is often plotted on a plan as a circle or other simple geometric shape. It can be used as a guide for establishing the specified TPZ.

specified tree protection zone (specified TPZ)—a TPZ that is adjusted in size or shape to accommodate the existing infrastructure, planned construction, and specific aspects of the site, as well as the tree canopy conformation, visible root orientation, size, condition, maturity, and species response to construction impacts.

When preserving groups of trees, and particularly when a new edge is cut through a stand, identify and consider removing trees that are tall with thin trunks and foliage concentrated at the top, suppressed or partially suppressed crowns, codominant stems, and other problems. The likelihood for failure of trees with these characteristics increases when neighboring trees are removed and the remaining trees are subject to greater wind exposure.

On larger sites, a temporary buffer composed of less valuable trees that will eventually be removed along the woodland edge could be used to temporarily protect those to be preserved. If this buffer is made up of small trees, they can be removed after construction activities but prior to landscaping. This buffer strategy can also be used when protecting an individual tree.

When groups of trees, such as buffer trees, are temporarily preserved, the TPZ should encompass the entire group and extend far enough away from the outer trees to protect roots and crowns.

Activities to Avoid Within the Tree Protection Zone (TPZ)

- Changing the grade (soil cuts and fills)
- Excavating trenches
- Cutting roots
- Pedestrian and equipment traffic that could compact the soil or physically damage roots
- Parking or operation of personal or construction vehicles or equipment
- Burning of brush or woody debris
- Storing soil, construction materials, petroleum products, water, or building refuse
- Disposing of wash water, fuel, or other potentially damaging liquids

Plan Review and Tree Impact Assessment

Development plans typically evolve from schematic drawings to design development (DD) to construction documents (CD) as work progresses through the planning, design, and construction phases, respectively. As the process proceeds, plans become more detailed and more difficult to modify.

Most project plans are prepared with computer-aided design (CAD) software programs to plot tree locations and create site plans. While individual project needs vary, the arborist's ability to effectively preserve trees may require accessing and reviewing plans in CAD, scalable PDF, or other formats.

Construction impacts are injuries or changes on a development site that affect tree health or structural stability. An arborist must be able to assess what those impacts will be before they occur and judge whether trees are likely to survive. Their ability to do so depends on their knowledge and experience with trees, skill at reading and interpreting development plans, and familiarity with construction. Being able to understand the planned construction and what injuries trees are likely to sustain based on those plans is a complex skill set that requires training and practice.

In many cases, the impacts to trees are readily assessed. Trees within areas where the structures and roads will be built (the building envelope) will be removed before construction. Trees far away from building areas can be preserved with minimal protection. The arborist should evaluate how the

Alternatives to Root Cutting

When root cutting is likely to severely damage trees, tree removal or alternative construction techniques should be considered. Typically, alternative construction techniques are more expensive, but arborists can work collaboratively with design teams on options for preserving trees. New techniques and technologies are emerging with the potential to minimize damage to trees, including:

- Moving construction farther from trees
- Tunneling under roots instead of trenching
- Bridging over roots
- Installing discontinuous footings for retaining walls, footings, and foundations
- Using flexible paving materials or shallower sections
- Reducing road, sidewalk, or pathway width
- Reinforcing with rods or underlaying with geotextile fabric
- Constructing supported/suspended pavement (e.g., supporting pavement on piers)

trees near the limits of construction will respond as the site is graded, utilities are installed, foundations are excavated, and buildings are constructed.

Grade Changes

Grade changes involve soil excavation (cuts) and/or addition of soil (fills) on a site to achieve a desired elevation. If the grade must be lowered near trees or excavation is required for other reasons, roots will likely be severed or damaged, potentially causing serious injury or instability. If roots within the CRZ are significantly impacted, a tree may be killed or destabilized. Arborists should evaluate the severity of the damage and how the tree is likely to respond.

Fill soils placed over roots and around trunks can alter gaseous (e.g., O_2 and CO_2) exchange and water movement, leading to tree decline and death. In particular, compaction of fill can damage roots and decrease soil water-holding capacity.

There are two types of fills. One is to raise the grade and/or dispose of excess soil, and the second is an engineered fill on which structures will be built. To construct an engineered fill that will support structures (e.g., buildings, sidewalks, and streets), the existing site soil is first scraped to remove unsuitable soil such as topsoil, and the exposed subgrade may then be compacted. The fill is then added in 6 to 12 in (15 to 30 cm) layers (lifts), and each layer is compacted. The degree and depth of compaction depends on the site's underlying geology and how much weight the soil must support. There are several ways to design the grade transition between the tree and the finish elevation. One way is to grade a slope from natural grade to the new grade, typically at 2:1 or 3:1 (run:rise) slope. Another way to transition the grade is to build a retaining wall (Figure 6). Constructing a retaining wall requires excavation to install a footing, so there is a trade-off between keeping the grade change

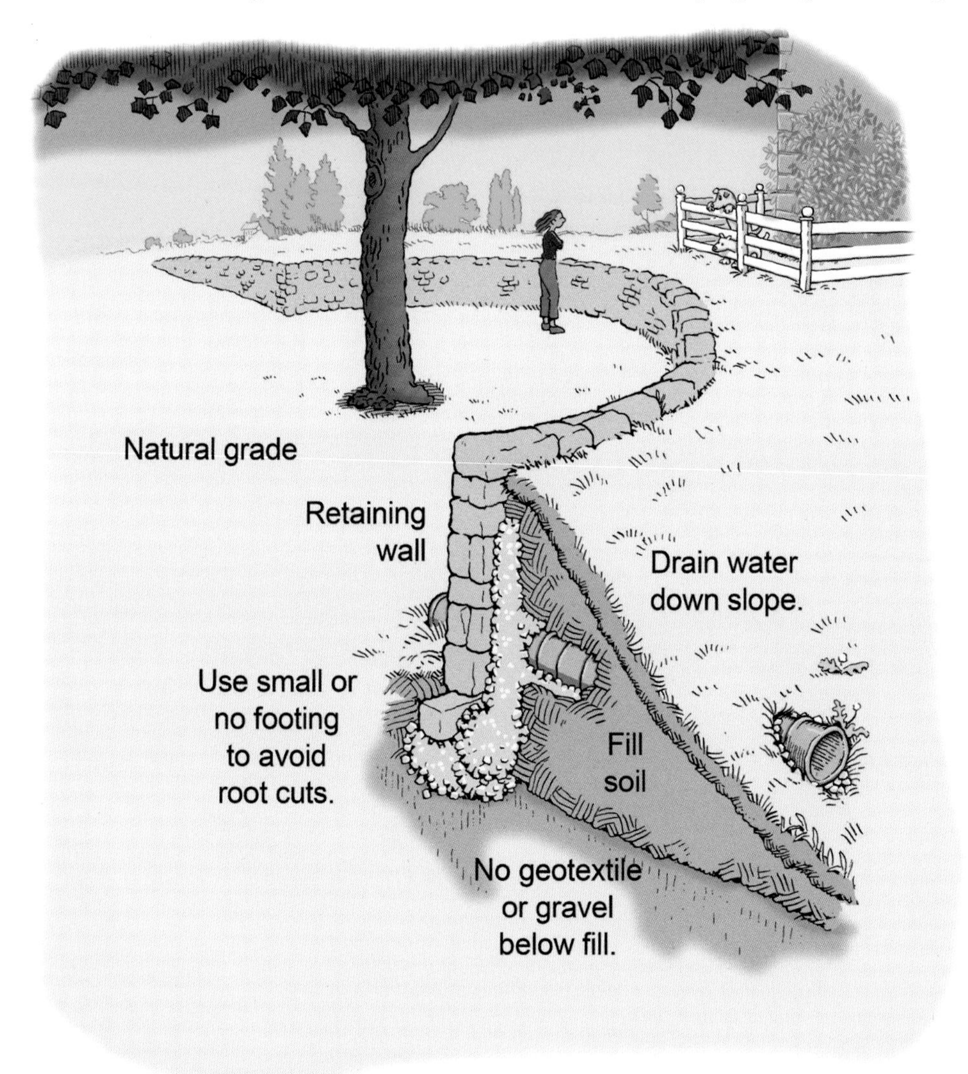

Figure 6. A retaining wall designed by a qualified professional may be constructed to minimize soil fill over tree root systems.

farther away from the tree and incurring root damage to excavate for a wall footing (Figure 7). A tree island may be needed when the grade is lowered completely around a tree (Figure 8).

When fill must be added around trees, soil aeration systems have sometimes been installed to promote oxygen and water movement into the original soil grade. There is no research that validates the effectiveness of aeration systems.

Changes to Water Flow

Site changes can affect water flow and soil hydrology. Diverting stormwater can increase or decrease soil moisture beyond the tolerance of existing plants even if this occurs far away from the tree (Figure 9).

Landscape Plan Review

Landscaping is often installed around trees that are preserved. Excavations within the TPZ to install landscape plants, lighting, and irrigation can cause extensive tree root damage. Arborists should review the plans and consider

Figure 7. A retaining wall may be used to minimize damage to roots as compared to cutting the entire slope. Retaining walls usually need to be permitted and designed by an engineer. If they are more than 3 ft (0.9 m) high, they may require a fence to protect people from falling.

Figure 8. A tree island may be needed when the grade is lowered completely around a tree. In this case, the tree's root system is permanently restricted and a system for applying supplemental water should be considered.

how these features impact trees. Check the depth and distance of excavations to the trunk to assess how much of the root system will be damaged. In most cases, it is best to keep underground conduit and irrigation lines well away from existing tree trunks and outside the specified TPZ.

When reviewing landscape plans, consider the tolerance of the trees to the additional irrigation that will be applied to the new plantings. Species to be planted within the TPZ should have water needs similar to the existing tree species to avoid over- or underirrigation.

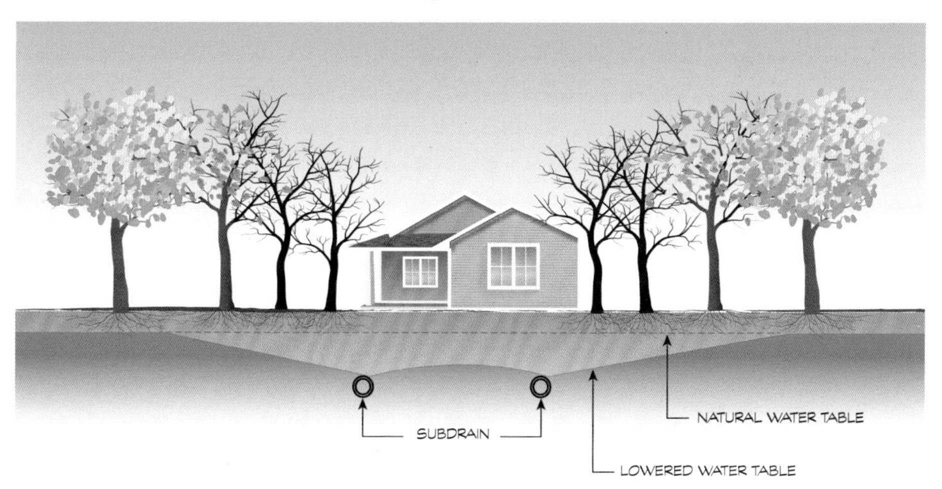

Figure 9. Changes in soil hydrology can affect trees at a different location near the construction site. In this illustration, installation of subdrains to lower the water table for construction has compromised the survival of retained trees that were dependent on the subsurface water.

Reporting During the Design Phase

During the design phase, the arborist should communicate with the development or design team. The arborist typically prepares a report that includes a list of trees and their location, condition, and suitability for preservation. In addition, it may include evaluation of impacts to trees, recommendations for tree preservation and removal, and specifications for tree protection and maintenance before, during, and after construction.

Tree Preservation Report

Tree preservation reports are usually required by the permitting agencies or the design team. Key elements of the report are often summarized on one or more plan sheets. Typically, the reports include:

- Project objectives
- Tree inventory data
- Map showing location of each tree
- Ratings of tree suitability for preservation
- Description of construction impacts to protected trees
- Disposition of each tree (preserve, transplant, or remove)
- Procedures and specifications for protecting trees, other plants, and soil areas
- Plan showing TPZs (if appropriate)
- Identification of permitting agency requirements
- Required mitigation for trees to be removed (if appropriate)
- Recommendations for care of trees (e.g., pruning to provide clearance for construction, pruning to improve structure, root pruning, soil management)
- Irrigation recommendations for trees during all phases
- Identification of the appropriate times to implement tree protection procedures. This may include a site monitoring schedule.
- Tree replacement requirements (if needed)
- Additional items as specified

The report should provide site-specific tree preservation guidelines. It should specify what activities are and are not allowed within the TPZ (e.g., no excavation and storage of equipment or materials), activities outside of the TPZ that should be avoided or reported (e.g., cutting roots larger than 2 in [5 cm] in diameter), and requirements for work to be reviewed or monitored by the arborist. Many permitting agencies have specific requirements for tree preservation; these should be incorporated into the report.

Portions of the report such as tree dispositions (remove or preserve) and specifications for tree protection measures and barriers should be included in the tree protection plan. This plan may be prepared by the arborist but is often drawn by others on the design team based on the information in the arborist's report. The tree protection plan is part of a larger package of development plans typically submitted to the permitting agency for review and comment along with the arborist's report (Figure 10). If approved, the plan set will direct the construction. The tree protection plan, as part of the development plans, instructs the construction team on how to protect the trees, work procedures needed, and who to call if problems arise.

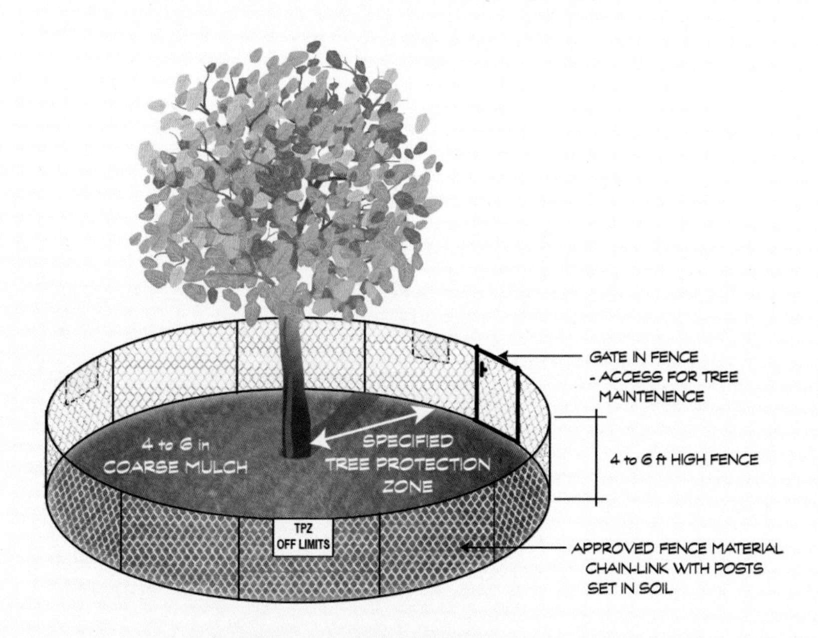

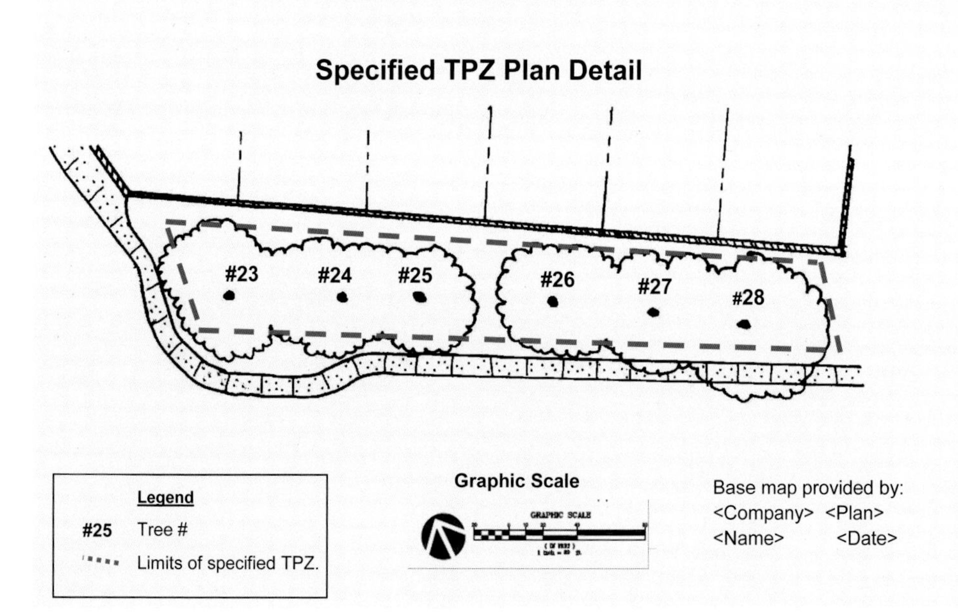

Figure 10. Example of a tree protection plan included in the development plan set. The information presented is derived from the tree protection report. (Adapted from Matheny and Clark 2023)

Tree Protection Notes

1. The specified tree protection zones (TPZs) are as plotted on the specified TPZ plan detail (below left).
2. Tree protection fences shall be installed at the limits of the specified TPZ before demolition begins and remain throughout construction.
3. Tree protection fences shall be 5 to 6 ft and chain-link. Fences are to be mounted on 2 in diameter galvanized iron posts, driven into the ground to a depth of at least 2 ft at no more than 10 ft spacing. A 3 ft wide opening shall be left in the fence to allow arborist access to the trees. See Tree Protection Zone detail.
4. No demolition, grading, excavation, construction, parking, storage of materials or spoil, or other activates may be conducted within the specified TPZ without prior approval and monitoring by the project arborist.
5. Refer to Tree Preservation Plan <author, date> for required tree protection and maintenance specifications during demolition, grading, and construction.
6. TPZ must be inspected and monitored by the project arborist throughout demolition and construction.
7. Refer to Tree Preservation Plan <author, date> for tree protection specifications during demolition and construction.

Tree Protection Plan

Project Name

Project Location

Prepared for:

Client/Owner

Prepared by:

Company

Address

Date

Tree Disposition Table

Tree #	Species	Trunk Diameter (inches)	Preserve/Remove
23	London plane	24"	Preserve
24	London plane	21"	Preserve
25	London plane	22"	Preserve
26	Raywood ash	16"	Preserve
27	Raywood ash	15"	Preserve
28	Raywood ash	18"	Preserve

Trees to Preserve = 6 Trees to Remove = 0

4. Preconstruction Phase

The preconstruction phase is the period between project approval (issuing of permits) and the beginning of construction. During this phase, site clearing, demolishing existing structures, and staking the layout of buildings, roads, and utility corridors will take place.

During the preconstruction phase, the arborist will begin to carry out or monitor the work identified in the tree protection plan. Tasks may include removing or transplanting trees, installing fences or other barriers at the limits of the specified TPZ, pruning roots and crowns that extend into the building footprint, installing temporary irrigation systems, mulching within the TPZ, and placing soil protection materials where needed.

This is the time for the arborist to meet with the developer's representative, site superintendent, and/or construction personnel to discuss work procedures adjacent to trees and what to do if problems arise. If construction personnel discover conflicts between trees and the approved construction, the arborist should work with the team to find practical solutions that protect trees while allowing construction to continue.

Prior to the beginning of work, hosting an educational session for construction personnel can raise awareness about the reasons to avoid damage to the soil and trees. This is a way to introduce the arborist to the construction personnel, build relationships, and potentially reduce the likelihood of inadvertent tree damage. Contractors should be made aware of the importance and requirements of the tree protection program, including penalties that will be assessed for unauthorized root cutting, soil compaction, trunk damage, branch removal, or other activities that may damage trees.

TPZ Barriers

Preconstruction tree work such as pruning, mulching, and installing temporary irrigation systems is often best accomplished before TPZ barriers are installed. Once that work is finished, barriers should be installed at the specified TPZ perimeter before any other site work is started, including demolition and site clearing.

Tree Protection Zone (TPZ) Barrier Specifications

Contracts for installation should include specifications for the:

- Type of barrier (e.g., chain-link fence, welded wire, wood fence, berms, buffer trees)
- Height of the barrier
- Method of anchoring the barrier in the ground
- Manner and timing of barrier installation
- Opening for access for tree inspection and maintenance
- Signage (size, type, information to be included, language(s), etc.)

TPZ barrier options include chain-link, welded wire, and wood fences; berms; buffer trees; and other devices. Fences should be sturdy and highly visible to discourage entry into the area. The fence should ideally be 4 to 8 ft (1.2 to 2.4 m) tall and solidly anchored into the ground. The preferred fence is chain-link, wire mesh, or wood. Plastic construction/snow fencing is easily moved or destroyed by construction activities and therefore is not recommended unless it is hung from a heavy wire attached between sturdy posts. In all cases, the fence should meet or exceed applicable permitting requirements. An opening may be left in the barrier to allow access to the tree for inspection and maintenance activities, such as mulch replenishment and irrigation.

TPZ barriers should be clearly marked with signs stating that the area within is being protected and that no one is allowed to enter or disturb this area without authorization (Figure 11). Signs should contain contact information for the contractor and/or arborist. Text on the signs should be in the language(s) commonly spoken on the site.

Figure 11. Example of tree protection zone (TPZ) signage.

Providing Temporary Construction Access

Foot or vehicular traffic and construction activities should be kept outside of the TPZ for the entire duration of construction. In some cases, there may need to be vehicle access within the TPZ. In these cases, trunk, branch, and/or soil should have temporary additional protection. Even with temporary protection, activities should be no closer than 5 ft (1.5 m) from the trunk.

Vertical Clearance

If temporary access occurs within the drip line, the distance to the lowest branch should be considered. The lowest branch needs to be higher than the equipment and its exhaust to avoid tree damage. Avoid temporary access routes that would require removal of large branches. Where needed and possible, it is better to tie branches out of the way rather than removing them.

Trunk Protection

Trunk protection is a temporary physical barrier installed to protect the trunk and/or buttress roots from mechanical damage when demolition or construction activities are expected to be close to the trunk (Figure 12). An example of these activities is removal of pavement close to a tree. One type of trunk protection is thick wood planks (dimensional lumber) around the trunk, preferably on a closed-cell foam or dimpled drainage board pad. Straps

Figure 12. Temporary trunk and buttress root protection structure for short-term activities, such as demolition near the trunk. This does not replace tree protection fencing to enclose the tree protection zone.

or wire are used to bind the planks in place. No fasteners should be driven into the tree. Wood barriers can be installed at an angle to protect the trunk flare and buttress roots. Another type of trunk protection is mesh tubes filled with straw or similar material (straw wattle) wrapped around the trunk.

When the work within the TPZ is completed, the trunk protection devices should be removed, and tree protection barriers should be reinstalled. Trunk protection devices left for too long can girdle or damage the tree. Straw wattle left for long periods of time can become waterlogged and damage the bark.

Soil and Root Protection

When additional soil or root protection is needed inside or outside the TPZ, actions can be taken to distribute the load, minimizing soil compaction and mechanical root damage (Figure 13). These treatments include:

- Applying 6 to 12 in (15 to 30 cm) of arborist wood chip mulch to the area

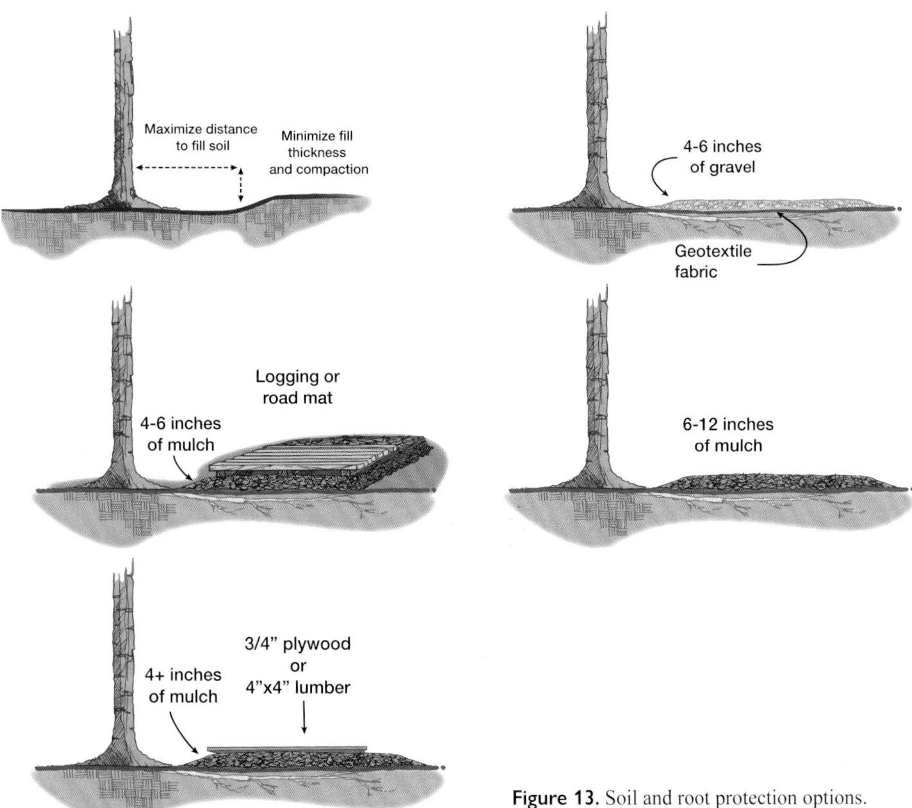

Figure 13. Soil and root protection options.

- Laying logging or road mat, load-bearing steel road plate, 0.75 in (2 cm) minimum thickness plywood, or beams over a 4+ in (10+ cm) thick layer of arborist wood chip mulch
- Applying 4 to 6 in (10 to 15 cm) of gravel with or without a cellular confinement web over a taut, staked, geotextile fabric. Gravel should not be applied directly over the soil surface because it will mix with the topsoil and be difficult to remove without damaging roots. (least preferred method)

Root protection materials will likely need to be removed at the end of the construction phase. Care must be taken to avoid scraping the soil and roots below during the removal operation. Wood chip mulches will break down and be incorporated into the soil if left on the soil surface; therefore, they do not need to be scraped off in many cases.

Preconstruction Arboricultural Treatments

Arboricultural treatments may be necessary to improve soil quality for roots within the TPZ or to improve tree health and structure. Mulch, irrigation, pest management, nutrient management, and pruning treatments should be applied as needed.

Some construction impacts can be mitigated with a planned tree health care program. This program may include recommendations for supplemental irrigation, organic mulching, and limitations on future uses and plantings within the rooting areas.

Preparing for Irrigation During Construction

Irrigation is one of the most important and often overlooked treatments on sites where roots are damaged, especially for drought-sensitive trees. Construction impacted trees may have a smaller soil water reservoir because of root removal, changes to impervious coverage, runoff/stormwater diversion, and/or soil compaction and manipulation. In addition, construction sites are often hotter and dryer than the site prior to development. Irrigation should be applied to compensate for the heat, loss of roots, and reduced soil volume.

If irrigation will be needed during construction, the arborist should determine how water will be conveyed and applied to trees. Irrigation application methods include aboveground sprinklers, bubblers, soaker hoses, drip systems, or injection of water into the soil. In some situations, a berm may be created, and flood irrigation provided. If drip irrigation lines or soaker hoses are installed, place the lines on the soil surface with mulch over the top.

Often, a stable source of water for irrigation is not available on a construction site. In this case, water will need to be brought to the site regularly or devices to store and distribute water will need to be installed.

Mulch

Coarse organic mulch applied to the soil surface is an effective way to conserve soil moisture, moderate soil temperatures, promote soil organisms, and protect the soil from compaction. Coarse mulch, such as fresh or partially composted arborist wood chips, should be maintained 2 to 4 in (5 to 10 cm) deep within the TPZ and in other areas to protect roots and soil. Mulch should never be placed against the trunk.

Plant Health Care

Foliar and/or soil samples should be collected and analyzed to determine nutrient status and fertilizer recommendations to satisfy a specific objective.[7] If nutrient deficiencies are negatively affecting tree health, fertilization before construction is recommended.

Certain tree species are known to have increased susceptibility to specific pests following construction stresses. Classic examples are wood-boring beetles, bark beetles, ambrosia beetles, cankers, and root decay fungi. Decay tends to spread more quickly in stressed trees. Low-vigor trees respond to wounds and pathogen invasion more slowly, and their capacity to resist fungal pathogens and close wounds is reduced. Some bark beetles and wood borers are attracted by volatile chemicals released by injured host trees.

In most situations, Integrated Pest Management (IPM) strategies should be employed.[8] Preventive treatments may be warranted prior to construction in regions where specific pests are known to be lethal. Trees on construction sites should be monitored regularly for indicators of pests, and action taken when problems arise.

---

7 See Smiley et al. 2020.

8 See Wiseman and Raupp 2016.

Pruning

Where construction will be near trees, the arborist should determine the amount of space needed for structures, equipment access, and operations, and should prune trees appropriately to provide clearance. In most cases, tree crowns should be pruned to provide 5 to 10 ft (1.5 to 3 m) clearance between the tree and structures (local requirements may vary). Pruning for clearance should be performed before equipment is brought in and work begins on the site.

Where temporary access is required, such as operating or moving tall equipment, it may be possible to temporarily tie back branches to provide clearance. Branch tiebacks should be removed when clearance is no longer needed.

Job Site Personal Protective Equipment

Most construction sites have strict requirements for personal protective equipment (PPE) that must be worn by all who are on the construction site. Typically, this includes:

- Hard hat
- Eye protection
- High-visibility clothing
- Long-sleeved shirt and long pants
- Work boots

Some work sites also require:

- Work gloves
- Respiratory protection
- Hearing protection
- Face shield

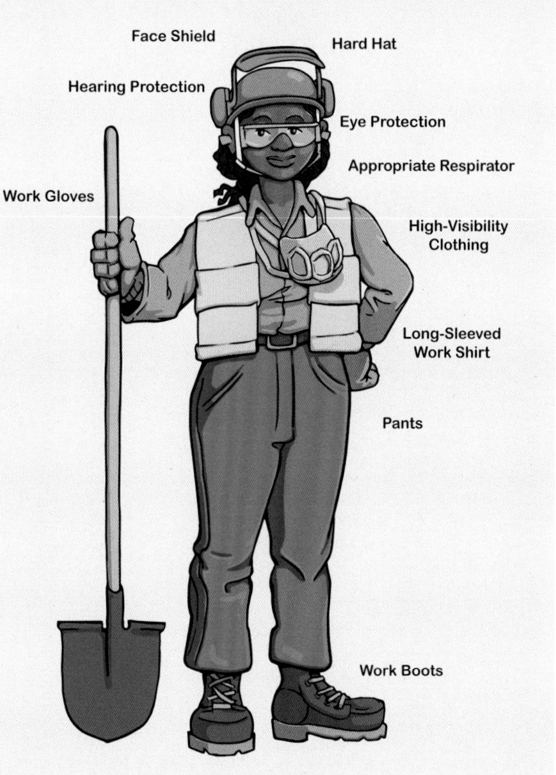

Find out what is required and wear the necessary equipment when on the work site, and always come prepared with the necessary PPE.

Pruning to compensate for root loss is seldom recommended. It is better to wait for the tree to respond to the root loss before determining what pruning is appropriate. If instability resulting from root loss is a concern, crown reduction or tree removal should be considered.

Trees also should be pruned to mitigate unacceptable risk to people, structures, or equipment in the construction area.[9] Supplemental supports and lightning protection systems should also be installed where they are needed and approved by the client.[10]

9 *See* Lilly et al. 2019.

10 *See* Smiley and Lilly 2014 and Smiley et al. 2015.

5. Construction Phase

The construction phase is when most activities occur at the development site. It starts with soil preparation and ends when construction has been completed. During the actual construction process, it is the arborist's role to stay in close contact with the contractor, developer, and permitting agency representative. The arborist should be an active, integral part of the construction team.

The arborist should inspect the site as requested by the client, required by the permitting agency, and specified in the scope of work. The main tasks are to maintain the integrity of the TPZ, monitor and report on the impact of work procedures, and look for and prevent tree damage. Periodic contractor meetings are recommended to keep the construction team informed about tree-related activities, and for the arborist to learn of construction-related activities and any design changes. Any tree damage or violations of contract agreements should be reported through a preestablished chain of command and remediated as soon as possible.

Arborists may find themselves involved in many ways during the construction phase. Typical roles include:

- Site monitoring—visiting the site to check on tree condition and tree protection, and monitor work close to trees
- Maintaining the tree protection zone—confirming the integrity of the TPZ and fencing, as well as responding to changes in construction plans that involve trees
- Root pruning—if excavation will occur within the root zone, cutting roots or monitoring the cutting of roots prior to soil removal operations
- Assessing impacts to trees—continuing to look at current and future planned construction to evaluate whether trees are likely to survive construction
- Responding to unforeseen conditions or conflicts that arise and providing solutions that protect the trees and facilitate progress

Assessing Impacts to Trees During Construction

Sometimes the arborist is not included in the project until construction has already started. When that occurs, the arborist must quickly learn the site design plans and review construction that has happened around trees. Unfortunately, there is usually little that can be done to change approved designs to reduce tree impacts, although slight modifications may be possible.

Even when the arborist has been part of the design team, evaluating construction impacts on trees continues through the construction phase. While on-site, the arborist should discuss planned work with the construction team and understand the work procedures. The arborist should assess how the trees will be impacted and, where appropriate, suggest changes to work procedures or design to better protect trees. The arborist should also recommend any additional treatments needed to mitigate the impacts. If the tree is likely to die or become unstable regardless of treatment, the arborist should recommend removing the tree.

Maintaining the Tree Protection Zone

Maintaining the TPZ is vital for successful tree preservation. The tree protection fencing placement is intended to adequately protect trees while allowing construction activities to go on outside it. Ideally, the construction crew never needs to enter the TPZ. During site visits, the arborist should inspect the TPZ, confirm that the tree protection barriers have not been relocated or damaged, and identify any necessary repairs or remedial action.

Despite efforts to avoid tree conflicts during the planning and design phases, changes sometimes occur, and approved plans sometimes include work within the TPZ. The arborist should meet with the construction personnel to understand the work required in the TPZ, recommend mitigating treatments, monitor the activity, and prepare a monitoring report. As examples, mitigating treatments could include root pruning, installing trunk or additional soil protection, or tree health care following completion of work within the TPZ.

Root Pruning

When soil is excavated near trees, roots may be pulled, broken, torn, or shattered some distance beyond the edge of the excavation. Root damage from excavation can cause great harm to a tree, especially if structural roots

How Construction Can Damage or Kill Trees

There are many ways that trees are damaged on construction sites; if the damage is extensive, the tree may be killed or become unstable. The common types of injury are listed below.

Root systems mechanically damaged by:

- Excavation equipment severing roots during grade changes or other excavation activities
- Trenching equipment used for gas, water, sewer, electrical, communication, irrigation, and other underground utility installations
- Installation of irrigation equipment and landscape plants

Root systems' growth and health reduced by:

- Soil compaction from vehicle and equipment traffic driving over the soil
- Fill soil added over the top of roots that alter water and air movement
- Soil or debris on the trunk or over the buttress roots that promotes the development of certain root disease pathogens and insect pests, as well as encourages stem girdling roots
- Alteration of hydrology of site impacting water supply to vegetation

Trunk or crown damaged by:

- Vehicle and machinery operations and access
- Heat from burning debris or equipment exhaust

Chemical damage from:

- Cleaning solvents, paint thinners, oils, and fuels
- Soil stabilizing treatments, such as lime
- Dust control with saline or contaminated water
- Contaminated water from concrete trucks, paint, and plaster clean outs
- Herbicide application

are affected. If pruning is performed at the edge of the planned excavation before soil is removed, damage to the remaining roots can be minimized.

To minimize tree damage, all roots over 1 in (2.5 cm) in diameter should be pruned cleanly rather than torn or crushed. Methods employed to prune roots include (Figure 14):

- Hand digging, using tools such as a shovel and pick to carefully remove the soil and leave roots intact. This method is slow and laborious, and it often results in inadvertent root damage.

- Soil excavation using air-excavation tools, pressurized water, vacuums, or hand tools, followed by selective root pruning.
- Cutting through the soil along a predetermined line on the surface using a tool specifically designed to cut roots. Even when root cutting machines are used, roots may need to be cleanly cut afterwards.

Air-excavation tools or high-pressure (greater than 100 psi [6.9 bars]) hydro-excavation tools can remove the soil around woody roots while causing very little damage. With either air or hydro excavation, if damage to the bark is seen, the nozzle should be moved away from the roots. During hydro excavation, a large vacuum system is usually needed to remove excess water and soil from the site.

Air, water, or hand excavation of the soil around the roots prior to root pruning allows the arborist to examine the roots and determine the best places to make selective pruning cuts, preferably beyond sinker roots or

Figure 14. Linear root cutting with trencher compared to selective root pruning after exposing roots with high-pressure air excavation.

Evaluating Root Loss Close to Tree Trunk

When root cuts are necessary, they should be located as far from the trunk as possible. Tree response to root cuts is dependent on tree species, age, condition, and root system configuration, as well as soil characteristics.

Cutting roots closer than six times the DBH on one side of the tree can cause sustained and chronic water-stress symptoms in some species. This stress in turn can lead to other tree health problems, such as increased susceptibility to pests, diseases, drought, or other environmental pressures.

When cuts are made closer to the trunk, stability and health may be compromised and should therefore be avoided. Immediate tree stability has been found to be compromised on some species when cuts are made at a distance from the trunk that is within three times the DBH.

For most species, when roots are cut at a distance from the trunk that is closer than one to one-and-ahalf times the DBH, immediate stability will be reduced, and long-term health and survival will be impacted. If large roots close to the trunk are to be cut, it may be better to remove the tree. If the tree is preserved, monitor the structural stability and minimize targets around the tree.

outside root branch unions. These methods also can be used to expose roots so that pipes or wires can be installed under or around them and so that walls or foundation posts can be strategically positioned between them.

Mechanical cutting tools such as root pruners, vibratory plows, or rock cutters are faster than digging by hand but may break, tear, or shatter larger roots. Mechanical root cutting is preferred over other forms of mechanical excavation such as a trencher or backhoe. Root cutting can also be used in combination with mechanical excavation by cutting the roots on the tree side of the excavation line prior to excavation so that the roots are not torn.

Root cutting using a trencher, excavator, or backhoe are the least preferred methods of root cutting. These machines are known to tear or crush roots at the excavation site and closer to the tree. If this type of root damage occurs, it is better to prune the damaged root ends to a flat surface with the adjacent bark firmly attached.

There are no wound-dressing treatments that have been shown to reduce root infection or decay.

Alternative to Trenching

Horizontal directional drilling and boring machines that tunnel under root systems to allow the installation of pipes, fiber optics, and/or wires without root severance may be alternatives to trenching. The boring equipment launch-and-recovery pits should be located outside the TPZ. In addition, the bore hole should be offset from the trunk by a distance of at least three times the DBH, measured from the face of the trunk (Figure 15).

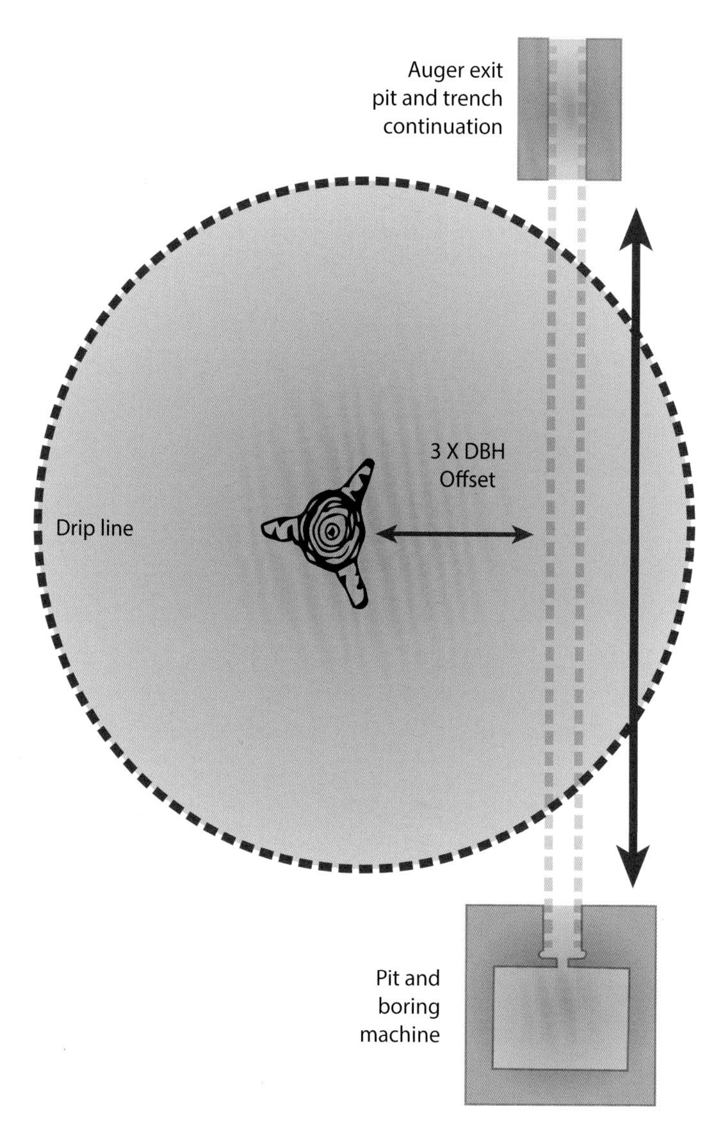

Figure 15. Boring under a tree root system is an alternative to trenching through the root system. The bore hole should be offset by a factor of three times the DBH from the trunk.

Irrigation During Construction

The need for irrigation, and the design and installation of the delivery system, was discussed in Section 4. Generally, irrigation should be applied within the TPZ of trees that are subject to water stress resulting from heat, loss of roots, or reduced soil volume.

Tree water needs vary with the season, species requirements, and degree of site changes. The frequency and volume of irrigation should be adjusted to meet the water demands of individual trees. Irrigation water should wet the soil to the depth of the tree roots that absorb water, generally the upper 6 to 18 in (15 to 45 cm) of the original soil surface. If spray irrigation is used, adjust the run time to ensure water adequately wets the mulch and the soil below.

To determine when to irrigate, the arborist should monitor soil moisture with sensors or by routinely extracting a soil sample from the root area and estimating by feel.[11] Allow enough time between irrigations for the soil to become slightly to moderately dry. Avoid frequent irrigation that maintains soil in a wet condition.

Site Monitoring

Site visits are necessary for the arborist to assess project compliance with the tree protection plan and integrity of the TPZ. Monitoring tree health and damage should be a priority during site visits. Special attention should be given to tree damage, soil moisture/irrigation, leaf wilting, unexpected changes in leaf color, premature leaf drop, branch dieback, canopy density, and the presence of primary and secondary pests. Broken branches should be recommended for pruning, and bark wounds should be treated as necessary (see Treating Bark Wounds sidebar).

Monitoring frequency should be defined in the project scope of work. In some cases, the permitting agency may define monitoring frequency. Monitoring events are best timed when the construction activity near trees takes place, so communication with the construction team is important for scheduling site visits.

11 See Scharenbroch and Smiley 2021.

Monitoring Report

The arborist should maintain records of site and tree observations as well as recommendations for tree protection or care. If requested, site monitoring records should be summarized in a report for the client. Documentation of inspections should be done as described in the contract or scope of work. Some permitting agencies have specific requirements for construction monitoring and reporting that should be followed. When possible, the arborist should discuss these observations and recommendations with the construction crew while on-site.

Treating Bark Wounds

If bark is damaged on the trunk or major branch, it may be possible to reattach the bark or wrap the area to encourage bark regeneration. Wound wrapping may not work on all species or at all times of the year, and it is only effective if done shortly after the damage occurs. With certain tree species, including many oaks (*Quercus* spp.), bark may regenerate if the damage does not penetrate deep into the xylem. To encourage this process, the trunk wound is wrapped with several layers of burlap or other semipermeable cloth to reduce drying. During the growing season, wraps should be left on for several weeks. Plastic wraps should not be used if the trunk wound is exposed to sunlight, as this may trap excessive heat.

Loose bark that is not reattached should be carefully removed or tacked back in place, leaving the attached bark intact. This does not need to be done immediately after wounding; the damaged bark is often left for the postconstruction phase. Jagged bark edges can be cut away with a sharp knife or chisel, taking care not to cut into living tissues or to make the wound wider. Cutting the perimeter of the wound into a smooth oval shape (bark tracing) has been shown to be of little or no value. Leaving peninsulas of live bark in a wound can speed closure; therefore, it is preferable to leave as much firmly attached live bark as possible.

Covering wounds with dressings (pruning paint, shellac, or latex paint) is not recommended unless it is known that open wounds attract serious insect pests to the tree species in the geographic area. Examples of these insects include the vectors of oak wilt fungus (*Bretziella fagacearum*) and pine bark beetles (*Dendroctonus* spp.). In these and other cases, treatments are needed during certain times of the year and should be applied soon after the cut is made or damage occurs.

6. The Landscaping Phase

Fine grading, installing irrigation, landscape lighting, planting, and other landscape features make up the last phase of a construction project. For the crew to accomplish this work, tree protection fences usually need to be adjusted or removed. Unfortunately, trees are often damaged during this phase.

During landscape installation, trees should be protected even when the barriers have been adjusted or removed; therefore, tree protection measures and monitoring should remain in place throughout the landscaping process if possible. As tree protection barriers come down for landscaping, arborist guidance to minimize trunk and root damage should be provided.

Soil is often rototilled in preparation for planting. This process can be very damaging to tree roots and should be minimized or prohibited within the TPZ. If the soil within the TPZ is compacted, alternative methods of soil preparation, such as with the use of air-excavation tools, should be used to minimize root disturbance.

Arborist Activities During the Landscaping Phase

- Continue to implement the tree protection plan
- Monitor the site on the specified schedule
- Work with contractors to minimize tree damage during landscape planting, hardscaping, and installation of irrigation, drainage, and landscape lighting
- Direct the adjustment of TPZ barriers to allow approved work
- Recommend or implement mitigation of tree damage

7. Postconstruction Phase

The postconstruction phase begins when the construction has ceased and landscape installation is finished. Once construction and landscaping are complete, the TPZ fencing may be removed, but monitoring by an arborist should continue if requested by the developer/owner or required by the permitting agency. Monitoring should include assessing soil moisture, mulch thickness, and tree damage, as well as inspecting for disease and insect pests. Treatments should be prescribed when problems are detected. Fertilization following construction should be based on soil and/or foliar analysis and use slow-release nutrient sources.

Tree health and structure should be evaluated again to determine if changes have occurred during the construction process. If changes are detected, mitigation treatments should be recommended and implemented. Pruning of live branches to compensate for root loss should not be performed because it may unnecessarily reduce tree vigor.

To help the property owner continue with an appropriate level of tree care, the arborist could prepare a long-term tree and landscape management plan, if asked to do so. This plan should outline a long-term management program that includes steps to sustain tree health. The report can include recommendations for monitoring, irrigating, pruning, fertilizing, managing pests, and scheduling maintenance, as needed.

Often commercial landscapes are densely planted. As they grow, there may be conflicts with trees that should be managed. If newly installed trees are staked or have temporary bags for irrigation, plans should be made for removal of these products.

Arborist Activities During the Postconstruction Phase

- Continue to implement the tree protection plan
- Monitor the site on the specified schedule
- Remove TPZ barriers
- Recommend or implement mitigation for tree damage
- Develop a maintenance plan for retained and newly installed plants, if included in the scope of work; include information on the need and standards for:
    - o Pruning
    - o Soil nutrient and compaction management
    - o Maintenance and removal of tree support systems
    - o Lightning protection
    - o Root management

Appendix

Developing Contracts for Tree Preservation

The arborist's scope of work should be defined by the type of project, the needs of the developer or owner, as well as any jurisdictional requirements.

The arborist and client should agree on the scope of work, including the following:

- Phases of development in which the arborist will be involved
- Attendance at meetings that affect trees
- Arborist responsibilities at each phase
- Location/area to be included
- Extent of tree and site evaluations
- Schedule of visits
- Type, timing, content, and recipient of reports

When contracting an arborist resource evaluation, specifications should include the following:

- Area to be evaluated
- Type of report to be provided (e.g., oral, written, map, plan sheet)
- Who will incorporate tree information onto a plan sheet
- Information included in the report
- Time frame for data collection
- Due date of the report
- To whom the report should be submitted

Glossary

aeration system—an underground arrangement of porous tubing installed in a tree or other plant's root area to improve gas exchange with the atmosphere. There is no research that demonstrates their effectiveness.

air excavator—device that directs a jet of highly compressed air to excavate soil. It is used to minimize damage to tree roots or underground structures, such as pipes and wires.

alternative building methods—construction techniques that reduce damage to trees compared to conventional designs. These techniques can include boring beneath roots, cantilevered building, grade beams, and foundation bridging.

calculated tree protection zone (calculated TPZ)—a TPZ that is calculated using the trunk diameter and a multiplication factor based on the species tolerance to construction and age of the tree. It is often plotted on a plan as a circle or other simple geometric shape. It can be used as a guide for establishing the specified TPZ.

codominant stems—forked branches nearly the same size in diameter, arising from a common union.

construction impacts—direct and indirect tree injuries or site changes that occur on a construction site that may affect the tree's health or structural stability.

construction phase—a stage in the development process when most activities occur. It starts with soil preparation and ends when construction has been completed.

critical root zone (CRZ)—area of soil around a tree where the minimum amount of roots considered critical to the health of the tree or structural stability are located. There are no universally accepted methods to calculate the CRZ.

DBH—acronym for tree trunk diameter at breast height: 4.5 ft above grade in the United States, 1.3 to 1.4 m in most other countries.

decay—1) n. a substance undergoing decomposition; 2) v. process of degradation by microorganisms.

design phase—a stage of the development process in which the location, size, and shape of the proposed buildings, landscape, utilities, and other infrastructure are decided and plotted on a site plan.

development—the process of converting land use for the installation of infrastructure or structures or making other site improvements.

grading—remodeling the landform by cutting and filling soil to change the surface elevation.

live crown ratio (LCR)—ratio of the height of the crown containing live foliage to the overall height of the tree.

permitting agency—governmental or nongovernmental organization that is responsible for issuing permits for development or tree work within a defined geographic area. This is often a city or county agency.

planning phase—the first stage of the development process in which the property owner and developer begin to assemble a team to explore potential land uses and to consider possible site layouts.

postconstruction phase—the last stage in the development process, which begins when the construction has ceased and landscape installation is finished.

preconstruction phase—the stage of the development process between project approval (issuing of permits) and the beginning of construction. During this phase, site clearing, demolishing existing structures, and staking the layout of buildings, roads, and utility corridors will take place.

preservation—the process of protecting trees and other vegetation, including the soil in which they grow, from damage related to development and construction activity as well as planning for their long-term health and stability.

preserved tree—a tree that is or was protected from development- and construction-related damage.

protected tree—a tree that, because of its size, species, age, or other qualities, requires permission from the permitting agency to remove, damage, prune, or otherwise impact.

pruning—removing branches or roots from a tree or other plant using approved practices to achieve a specified objective.

resource evaluation—the process of assessing the trees or other vegetation, and/or soil, present on or near the site, including determining legal requirements, potential challenges, and opportunities.

root cutting—severing of roots nonselectively.

root pruning—severing roots selectively. The process of cutting roots at the line of a planned excavation to prevent tearing and splintering of remaining roots during grading or excavation.

soil compaction—compression of the soil, often as a result of vehicle or heavy-equipment traffic, that breaks down soil aggregates and reduces soil volume and total pore space, especially macropore space.

specifications—detailed plans, requirements, and statements of procedures and/or standards used to define and guide work.

specified tree protection zone (specified TPZ)—a TPZ that is adjusted in size or shape to accommodate the existing infrastructure, planned construction, and specific aspects of the site, as well as the tree canopy conformation, visible root orientation, size, condition, maturity, and species response to construction impacts.

supported/suspended pavement—paving option that transfers the load to the subsoil rather than the top of the soil surface.

tree—a woody perennial plant with a single or multiple trunks which typically develop a mature size of over several inches (centimeters) in diameter and 10 or more feet (3 m) in height.

tree inventory—a map, plan, and/or list that describes tree attributes, such as location, species, size, and/or condition.

tree preservation (in the United Kingdom, tree retention)—process of protecting trees, shrubs, other vegetation, and soil from damage related to development and construction activity, as well as planning for their long-term health and stability.

tree protection—process of excluding activities that could cause damage to a tree or soil that would negatively affect tree health or longevity.

tree protection plan—a component of the construction plan set showing trees to be preserved, specified tree protection zones, work procedures, maintenance, and monitoring requirements for protecting trees, shrubs, other vegetation, and soil during construction.

tree protection zone (TPZ)—area within which certain activities are prohibited or restricted to prevent or minimize potential injury to designated trees, especially during construction or development. The TPZ should encompass the critical root zone, based on the judgment of the arborist.

tree risk assessment—a systematic process used to identify, analyze, and evaluate tree risk.

trenching—narrow excavation, often used to install utilities or structural footings.

tunneling—digging, often with special machinery, below the surface of the ground without an open trench.

vigor—capacity to grow and resist stress. Sometimes limited in reference to genetic capacity.

water table—upper level of groundwater in the soil.

Selected References and Other Sources of Information

American National Standards Institute. 2019. *American National Standard for Tree Care Operations—Tree, Shrub, and Other Woody Plant Management—Standard Practices (Management of Trees and Shrubs During Site Planning, Site Development, and Construction)*(ANSI A300, Part 5). Manchester (NH, USA): Tree Care Industry Association, Inc. 21 p.

Benson A, Koeser A, Morgenroth J. 2019. A test of tree protection zones: responses of live oak (*Quercus virginiana* Mill) trees to root severance treatments. *Urban Forestry & Urban Greening*. 38:54–63.

Benson A, Morgenroth J, Koeser A. 2019. Responses of mature roadside trees to root severance treatments. *Urban Forestry & Urban Greening*. 46:126448.

Bond J. 2013. *Tree Inventories*. 2nd Ed. Best Management Practices. Champaign (IL, USA): International Society of Arboriculture. 35 p.

British Standards Institute. 2012. BS 5837:2012. *Trees in Relation to Design, Demolition and Construction—Recommendations*. London (UK): British Standards Institute. 42 p.

Coder KD. 1995. Tree quality BMPs for developing wooded areas and protecting residual trees. In: Watson GW, Neely D, editors. *Trees and Building Sites*. Savoy (IL, USA): International Society of Arboriculture.

Costello LR, Watson G, Smiley ET. 2017. *Root Management*. 1st Ed. Best Management Practices. Champaign (IL, USA): International Society of Arboriculture. 41 p.

Council of Tree and Landscape Appraisers. 2019. *Guide for Plant Appraisal*. 10th Ed. Atlanta (GA, USA): International Society of Arboriculture. 181 p.

Day SD, Wiseman PE, Dickinson SB, Harris JR. 2010. Contemporary concepts of root system architecture of urban trees. *Arboriculture & Urban Forestry*. 36(4):149–159.

Elmendorf W, Gerhold H, Kuhns L. 2005. A guide to preserving trees in development projects. University Park (PA, USA): Penn State College of Agricultural Sciences. 25 p. https://extension.psu.edu/a-guide-to-preserving-trees-in-development-projects

Harris RW, Clark JR, Matheny NP. 2004. *Arboriculture: Integrated Management of Landscape Trees, Shrubs, and Vines*. 4th Ed. Upper Saddle River (NJ, USA): Prentice Hall. 687 p.

Johnson GR. 1999. Protecting trees from construction damage: a home-owner's guide. St. Paul (MN, USA): Minnesota Digital Conservancy. https://hdl.handle.net/11299/199785

Lilly SJ, Gilman EF, Smiley ET. 2019. *Pruning*. 3rd Ed. Best Management Practices. Atlanta (GA, USA): International Society of Arboriculture. 63 p.

Matheny NP, Clark JR. 1998. *Trees and Development: A Technical Guide to Preservation of Trees During Land Development*. Champaign (IL, USA): International Society of Arboriculture. 184 p.

Morell JD. 1988. Utility and municipal communications relating to the urban forest. *Journal of Arboriculture*. 14(11):273–275.

Roman LA, van Doorn NS, McPherson EG, Scharenbroch BC, Henning JG, Östberg JPA, Mueller LS, Koeser AK, Mills JR, Hallett RA, Sanders JE, Battles JJ, Boyer DJ, Fristensky JP, Mincey SK, Peper PJ, Vogt J. 2020. Urban tree monitoring: a field guide. Madison (WI, USA): US Department of Agriculture, Forest Service, Northern Research Station. General Technical Report No. NRS-194. 48 p.

Scharenbroch BC, Smiley ET. 2021. *Soil Management for Urban Trees*. 2nd Ed. Best Management Practices. Champaign (IL, USA): International Society of Arboriculture. 70 p.

Smiley ET, Graham AW Jr, Cullen S. 2015. *Tree Lightning Protection Systems*. 3rd Ed. Best Management Practices. Champaign (IL, USA): International Society of Arboriculture. 61 p.

Smiley ET, Lilly S. 2014. *Tree Support Systems: Cabling, Bracing, Guying, and Propping*. 3rd Ed. Best Management Practices. Champaign (IL, USA): International Society of Arboriculture. 50 p.

Smiley ET, Matheny N, Lilly S. 2017. *Tree Risk Assessment*. 2nd Ed. Best Management Practices. Champaign (IL, USA): International Society of Arboriculture. 86 p.

Smiley ET, Werner L, Lilly SJ, Brantley B. 2020. *Tree and Shrub Fertilization*. 4th Ed. Best Management Practices. Atlanta (GA, USA): International Society of Arboriculture. 57 p.

Urban J. 2008. *Up By Roots*. Champaign (IL, USA): International Society of Arboriculture. 479 p.

Wiseman PE, Raupp MJ. 2016. *Integrated Pest Management*. 2nd Ed. Best Management Practices. Champaign (IL, USA): International Society of Arboriculture.

About the Authors

Nelda Matheny is a senior consulting arborist with HortScience | Bartlett Consulting. She is a coauthor of several books, including *Trees and Development* and *Arboriculture: Integrated Management of Landscape Trees, Shrubs, and Vines.* She has 40 years of experience preserving trees on development projects.

E. Thomas Smiley, PhD, is a senior arboricultural researcher at the Bartlett Tree Research Laboratories. He serves on the ANSI A300 Standards for Tree Care Operations Committee and has authored many ISA publications. He received his PhD from Michigan State University, MS from Colorado State University, and BS from the University of Wisconsin–Madison.

Ryan Gilpin is principal consultant of Nidus Consulting based in Portland, OR. Ryan has been a consulting arborist working on tree preservation projects since 2011 and coauthored the *Tree Care for Wildlife Best Management Practices.* Ryan has an MS from Georg-August-Universität and a BS from the University of California, Davis.

Richard Hauer, PhD, is a professor of urban and community forestry at the University of Wisconsin–Stevens Point. He serves on the ANSI A300 Standards for Tree Care Operations Committee. He received his PhD from the University of Minnesota, MS from the University of Illinois, and BS from the University of Wisconsin–Stevens Point.